Chlory the Green Pig Makes Sugar

A Food For Thought Book™

Written by Geoffrey Wilhite
Background art by Geoffrey Wilhite
Digital graphics and editing by
Happy Dolphin Press

Copyright ©2018 by Geoffrey Wilhite

All rights reserved. In accordance with the U.S. Copyright Act of 1976, the scanning, uploading, and electronic sharing of any part of this book without the permission of the publisher constitutes unlawful piracy and theft of the author's intellectual property. If you would like to use material from this book (other than for review purposes) prior written permission must be obtained by contacting the publisher at permissions@happydolphinpress.com. Thank you for your support of the author's rights.

Published by
Happy Dolphin Press
13550 Reflections Parkway
Suite 5-501
Fort Myers, FL 33907
www.HappyDolphinPress.com

Printed in the United States of America

First Edition April 2018

Happy Dolphin Press publishes children's authors and produce fun, educational books designed to grow with readers. Contact us at events@happydolphinpress.com to schedule an author event or speaker.

"We take fun seriously!" ™

Bulk quantities of this and all Happy Dolphin Press books are available at a discount through the publisher at sales@happydolphinpress.com

ISBN 13: 978-1-947678-06-4 (hardcover)
ISBN 10: 1-947678-06-X (hardcover)

ISBN 13: 978-1-947678-04-0 (paperback)
ISBN 10: 1-947678-04-3 (paperback)

LCCN: 2018903910

Digital Graphics and Editing was provided by Jennifer Smith through Happy Dolphin Press

Early in the morning, a sleepy sunflower is happy when she feels the warm sun because she produces her own food during a process called photosynthesis.

In her leaves there is a place called Chloroplast where little green pigs named Chlorophyll live.

It's their pigment that gives them this color.

All day long they are busy making sugar for the sunflower.

One of these pigs making sugar in Leaf Town is a famous chef named Chlory.

Her friends, Phloem and Xylem, are long skinny tubes who are helping her.

"Let's see," said Chlory, "what three ingredients do I need to make sugar?"

"I know," said Phloem, jumping up and down. "Sunshine. You need sunshine."

"Wonderful!" said Chlory. "Now for the second ingredient."

"I know, I know!" said Xylem, twirling around. "Water is the second ingredient."

"Correct again!" said Chlory, "And where do we get the water?"

"From me," said Xylem proudly.

Xylem is full of water, and it's his job to deliver the water and minerals from the soil into the roots. He then carries the mixture up to all parts of the sunflower.

"Now, one more thing," said Chlory. "What's in the balloons?"

"Oh, I know," said Phloem, waving her arms in the air. "Carbon Dioxide is in the balloons."

"True," said Xylem smirking, "but the cool tubes call it CO2!"

"And that's the third ingredient." said Chlory. "Now we can mix them all together and make sugar!"

"But look, the balloons are wilting!" cries Xylem. "Now we can't make any more sugar!"

"Oh no! There's not enough CO_2 in the balloons," said Phloem. "What are we going to do?"

"I know what to do," said Chlory.
"We'll go to Body Town to see my friend Mito, but first give me a few minutes to get ready."

Chlory puts on her best dress,
dolls up her face, and sets out to impress.
Then grasping the ballons, she says,
"Follow me!"

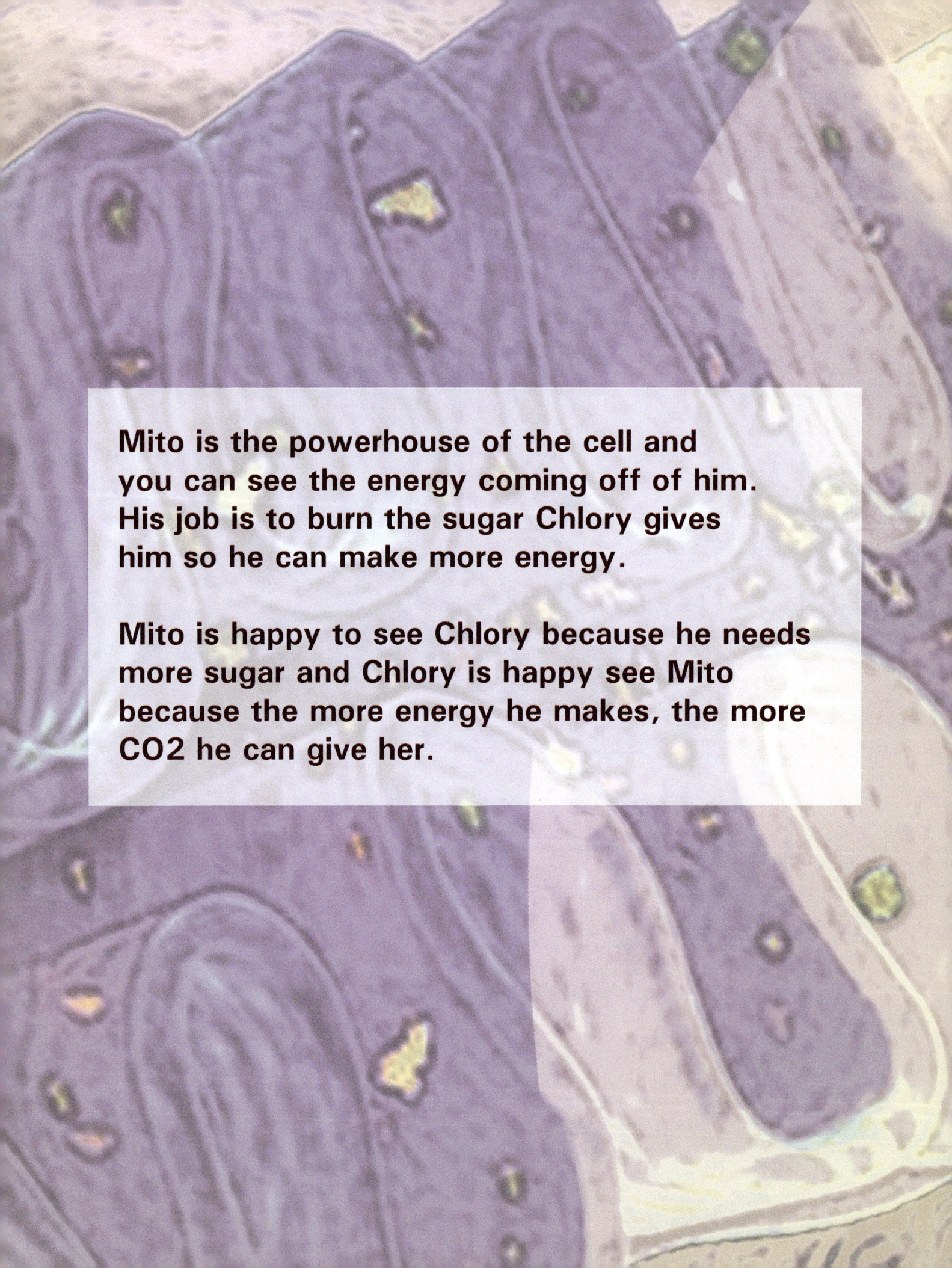

Mito is the powerhouse of the cell and you can see the energy coming off of him. His job is to burn the sugar Chlory gives him so he can make more energy.

Mito is happy to see Chlory because he needs more sugar and Chlory is happy see Mito because the more energy he makes, the more CO2 he can give her.

"I'll give you more sugar," Chlory says, "if you can give me more CO_2."

"Okay." says Mito.

Chlory gives Mito a great big hug filled with sugar. Mito turns that sugar into energy.

With that energy, he fills the balloons with CO_2.

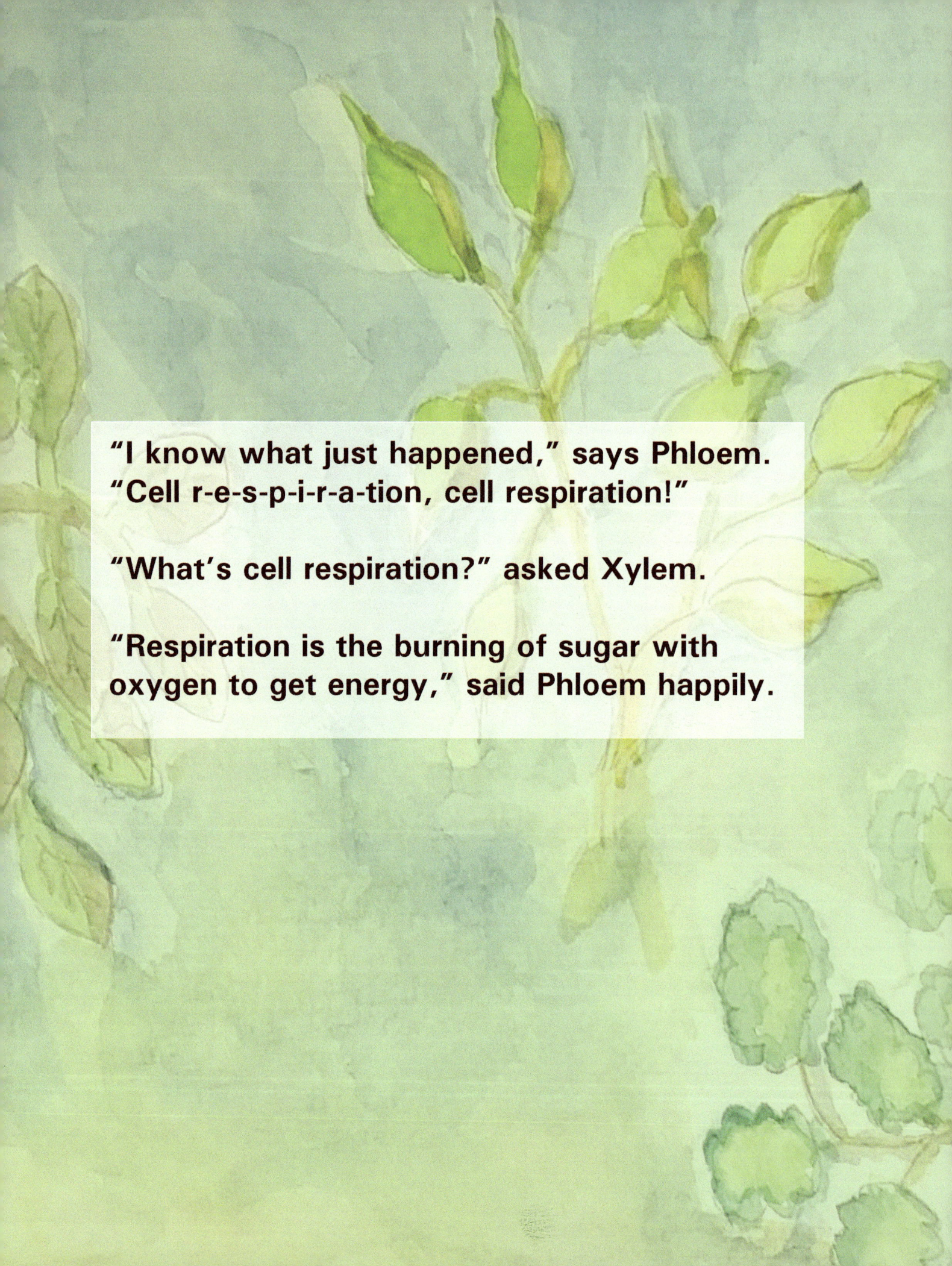

"I know what just happened," says Phloem. "Cell r-e-s-p-i-r-a-tion, cell respiration!"

"What's cell respiration?" asked Xylem.

"Respiration is the burning of sugar with oxygen to get energy," said Phloem happily.

Sugar!

Phloem takes some of the sugar and delivers it to all the parts of the sunflower plant.

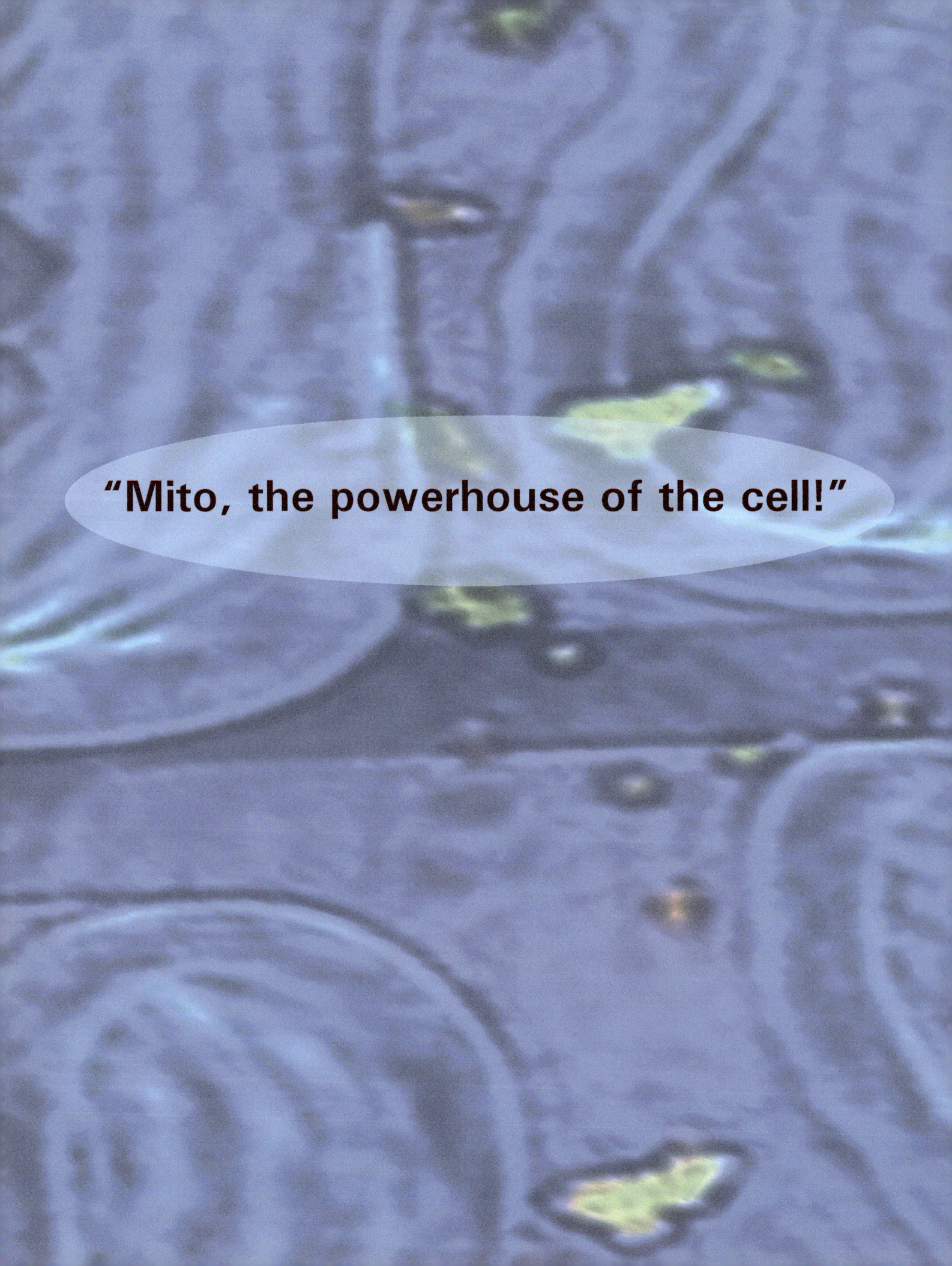

 # Glossary

Carbon dioxide (CO2): a colorless gas that is formed especially by the burning and breaking down of sugar (cell respiration). The CO2 is absorbed from the air by plants during photosynthesis.

Cell Respiration: the process by which cells use oxygen to produce energy from sugar

Cell: in biology, the smallest unit that can perform all life processes; cells are covered by a membrane and contain DNA and cytoplasm

Chlorophyll: a green pigment that captures light energy for photosynthesis

Chloroplast: an organelle that contains chlorophyll and is the site of photosynthesis

Leaf: one of the green usually flat parts that grow from a stem or twig of a plant and that function mainly in making food by photosynthesis

Minerals: a naturally occurring substance (as ore, petroleum, or water) usually obtained from the ground

Mitochondria: a cell organelle that is the site of cellular respiration

   I. **Organelle:** any of a number of organized or specialized structures within a living cell.

Oxygen: one of the main elements that make up air used to burn sugar for energy

Phloem: the tubes that conduct sugar in vascular plants

Photosynthesis: the process by which plants, use sunlight, carbon dioxide, and water to make food (sugar)

Pigment: a coloring organelle in animals and plants especially in a cell or tissue

Plant: photosynthetic living things usually lacking the ability to move from place to place under their own power, having no obvious nervous or sensory organs, possessing cellulose cell walls.

Producer: an organism that can make its own food by using energy from its surrounding

Roots: the leafless usually underground part of a plant that absorbs water and minerals, stores food, and holds the plant in place

Sugar (Glucose): occurring especially in a natural form that is found in plants, fruits, and is a source of energy for living things.

Vascular: the tissue and series of tubes and veins that move nutrients collected by the roots through the plant to the stem and leaves.

Veins: one of the vascular bundles forming the framework of a leaf

Water (H_2O): a fluid necessary for the life of most animals and plants

Wilting: to lose or cause to lose freshness and become limp

Xylem: tubes that transport water and minerals in vascular plants

In loving memory of Lucy.